SpringerBriefs in Fire

Series Editor
James A. Milke

For further volumes:
http://www.springer.com/series/10476

Code Consultants, Inc.

Antifreeze Solutions in Home Fire Sprinkler Systems

Code Consultants, Inc.
St. Louis, MO, USA

ISSN 2193-6595 ISSN 2193-6609 (electronic)
ISBN 978-1-4614-3839-7 ISBN 978-1-4614-3840-3 (eBook)
DOI 10.1007/978-1-4614-3840-3
Springer New York Heidelberg Dordrecht London

Library of Congress Control Number: 2012936070

Springer is part of Springer Science+Business Media (www.springer.com)

Preface

NFPA 13, *Standard for the Installation of Sprinkler Systems*, has included guidance on the use of antifreeze solutions in fire sprinkler systems since the 1940 edition. [1] Antifreeze solutions may be used in fire sprinkler systems where the piping system, or portions of the piping system, may be subject to freezing temperatures. [2] Antifreeze solutions permitted for use in sprinkler systems connected to potable water supplies include propylene glycol and glycerin.

Recent fire incidents, analysis of available literature, and preliminary testing have identified concerns with the use of certain antifreeze solutions. Under certain conditions, solutions of glycerin and propylene glycol antifreeze have been found to ignite when discharged from automatic sprinkler systems. A literature review, preliminary testing, and a long term research plan were developed as part of Phase I of this project. This Report outlines the results of Phase II of the project, which includes further testing of propylene glycol and glycerin antifreeze solutions for a range of concentrations and operating conditions. The testing and analysis were limited to antifreeze solutions discharged through residential sprinklers and did not investigate other types of sprinklers.

A test plan was developed for Phase II to investigate the potential for large-scale ignition of antifreeze solutions discharged from residential sprinklers and the influence of antifreeze solutions on the effectiveness of residential sprinkler systems in controlling a fire condition and maintaining tenable conditions for egress. Testing was conducted in two parts. Scope A consisted of fire tests using six (6) models of sprinklers at elevations of 8 ft and 20 ft to investigate the potential for large-scale ignition of antifreeze sprays at pressures ranging from 10 psi to 150 psi. Scope B consisted of room fire tests, similar to UL 1626, that were designed to investigate the effective of sprinklers discharging antifreeze solutions and their ability to maintain tenable conditions.

Results of the Scope A testing indicate that concentrations of propylene glycol exceeding 40% by volume and concentrations of glycerin exceeding 50% by volume have the potential to ignite when discharged through residential sprinklers. The potential for ignition depends on several factors including the ignition source, sprinkler model, sprinkler elevation, discharge pressure, and the location of the sprinkler with respect to the ignition source. Ignition of antifreeze spray increased the measured heat release rate in certain tests with 50% propylene glycol and 55% glycerin by more than 300%. For certain test conditions, the increase in heat release rate resulting from the application of 55% glycerin solution exceeded the increase in heat release rate from the application of 50% glycerin solution by a factor of 10. A similar level of sensitivity was observed between 40% and 50% propylene glycol solutions, but not between 40% and 45% propylene glycol solutions.

The results of the Scope B testing indicated that concentrations of propylene glycol not exceeding 40% by volume and concentrations of glycerin not exceeding 50% by volume have similar performance to water as compared to the UL 1626 fire control criteria. Both the 40% propylene glycol and 50% glycerin solutions met the UL 1626 fire control criteria and demonstrated similar performance to that of water alone throughout the series of tests.

The results of this research suggest that antifreeze solutions of propylene glycol exceeding 40% and glycerin exceeding 50% by volume are not appropriate for use in home fire sprinkler systems. Consideration should be given to an appropriate safety factor for concentrations of these antifreeze solutions that are permitted by future editions of NFPA 13, as well as warnings and limitations outlined in anti-freeze product literature.

Based on the flammability properties outlined in Table 4, the use of solutions of diethylene glycol and ethylene glycol in home fire sprinkler systems should also be limited unless testing is conducted to establish that they are appropriate for use in home fire sprinkler systems. The results of this analysis are limited to residential sprinklers; the flammability of antifreeze solutions discharged through other types of sprinklers has not been investigated.

Recommendations are provided for further research in the following areas:

- Investigate the use of Antifreeze Solutions in Sprinkler Systems with Non-Residential Sprinklers
- Characterize Droplet Size Distributions from Sprinklers
- Develop a Small or Medium Scale Screening Test of Antifreeze Solutions
- Develop a Listing Standard for Solutions introduced into Sprinkler Systems

This report is the second in a series of research reports published by the Foundation reporting on research on this topic.

Contents

I
Introduction

Recent fire incidents raised questions regarding the effectiveness of antifreeze sprinkler systems in controlling residential fire conditions. As a result, the Fire Protection Research Foundation retained Code Consultants, Inc. (CCI) to perform a literature search and develop a research plan to investigate the impact of antifreeze solutions on the effectiveness of home fire sprinkler systems. [3] The literature review included the following subjects:

1. Antifreeze sprinkler system requirements
2. Mixing and separation of antifreeze compounds commonly used in sprinkler systems
3. Flammability of antifreeze solutions commonly used in sprinkler systems
4. Factors influencing the flammability of liquids, such as dispersion in droplets
5. Characterization of residential sprinkler sprays
6. Factors influencing the potential for flash fires or explosions from liquid sprays
7. Existing fire test data on the effectiveness of antifreeze solutions at controlling fire conditions
8. Fire incident reports involving antifreeze sprinkler systems

A research plan was developed to supplement the literature review in areas where existing information was limited. In addition, CCI observed a series of preliminary fire tests (Phase I) conducted by Underwriters Laboratories, Inc. (UL) to investigate the effectiveness of antifreeze sprinkler systems in controlling certain home fire scenarios. A summary of data from the preliminary testing was also provided by UL. [4] CCI provided suggestions for further research to provide a more complete analysis of currently permitted antifreeze solutions as well as to investigate other antifreeze solutions that could be used in sprinkler systems.

The Phase I testing identified the need for additional research regarding the potential for antifreeze solutions to create a large-scale ignition of spray when discharged through automatic sprinklers onto a fire. Additional research was also needed to further investigate the impact of antifreeze solutions on a sprinkler system's ability

Code Consultants, Inc., *Antifreeze Solutions in Home Fire Sprinkler Systems*,
SpringerBriefs in Fire, DOI 10.1007/978-1-4614-3840-3_1,
© Fire Protection Research Foundation 2010

to control a fire condition and maintain tenable conditions. As such, a Phase II test plan was created based on the Phase I information.

The Phase II test plan was separated into two scopes (A and B) which were intended to investigate additional research identified in Phase I. Scope A tested antifreeze solutions for the potential to create a large-scale ignition of the spray when discharged through sprinklers onto a fire. Scope B tested antifreeze solutions for its impact on a sprinkler system's ability to control a fire condition and maintain tenable conditions.

II
Background

Recent fire tests have indicated the potential for ignition of certain antifreeze solutions discharged from automatic sprinkler systems. [4] The potential for ignition of an antifreeze spray and the influence of antifreeze solutions on sprinkler effectiveness involves several complex and contemporary research topics. This section provides a basic summary of relevant background information; a more complete discussion can be found in the report from Phase I of this project. [3]

A   Antifreeze Solutions

NFPA 13 [5], 13D [6], and 13R [7] each include substantially similar requirements for antifreeze solutions used in sprinkler systems. Antifreeze solutions of propylene glycol and glycerin with water are each permitted in sprinkler systems connected to potable water supplies. The antifreeze solutions are intended to protect sprinkler piping that passes through areas that could be exposed to freezing temperatures. Freezing point data for propylene glycol and glycerin solutions provided in NFPA 13 is summarized in the following table.

Table 1 Adapted from NFPA 13 Table 7.6.2.2 Antifreeze Solution to be Used if Potable Water is Connected to Sprinklers

Material	Solution with Water (by Volume)	Specific Gravity at 60 °F (15.6 °C)	Freezing Point °F	°C
Glycerin (C.P. or U.S.P grade)	50% glycerin	1.145	−20.9	−29.4
	60% glycerin	1.171	−47.3	−44.1
	70% glycerin	1.197	−22.2	−30.1
Propylene glycol	40% propylene glycol	1.034	−6	−21.1
	50% propylene glycol	1.041	−26	−32.2
	60% propylene glycol	1.045	−60	−51.1

C.P.: Chemically pure. U.S.P.: United States Pharmacopoeia 96.5%

Code Consultants, Inc., *Antifreeze Solutions in Home Fire Sprinkler Systems*,
SpringerBriefs in Fire, DOI 10.1007/978-1-4614-3840-3_2,
© Fire Protection Research Foundation 2010

Antifreeze solutions of ethylene glycol and diethylene glycol are also permitted, but only in sprinkler systems that are connected to non-potable water supplies. This research focuses on propylene glycol and glycerin antifreeze solutions, because they are believed to be much more commonly used in home fire sprinkler systems.

Antifreeze solutions of glycerin, diethylene glycol, and ethylene glycol were included in NFPA 13 starting in the Appendix of the 1940 edition, known at the time as National Board of Fire Underwriters Pamphlet No. 13. [1] The 1953 edition of NFPA 13 included requirements for antifreeze sprinkler systems in the body of the standard and permitted the use of propylene glycol or calcium chloride solutions as well as glycerin, diethylene glycol, and ethylene glycol. [8] The antifreeze solutions and concentrations permitted by the 1953 edition of NFPA 13 are the same as those permitted by the current (2010) edition of NFPA 13, with the exception that calcium chloride is no longer permitted. [2]

Table 2, below, illustrates the freezing point and the specific gravity values (at 25°C) for a propylene glycol-water mixture in addition to the corresponding percent volume and percent weight of propylene glycol. The difference in percent volume and percent weight of propylene glycol solutions is minimal, because its density is only slightly higher than the density of water.

Table 2 Propylene Glycol Properties [9] (Portions of data are calculated or interpolated)

Propylene Glycol Properties			
%Vol.	%Wt.*	Freezing Point (°F)	Specific Gravity at 25 °C
0%	0%	32	1.000
5%	5%	26	1.004
10%	10%	25	1.008
15%	15%	22	1.012
20%	20%	19	1.016
25%	25%	15	1.020
30%	30%	11	1.024
35%	35%	2	1.028
40%	40%	−6	1.032
45%	45%	−18	1.035
50%	50%	−26	1.038
55%	55%	−45	1.040
60%	60%	−60	1.041

*% Vol. to % wt. conversion is at 25 °C

Similar to Table 2 (above), Table 3 (below) depicts the freezing point and specific gravity values (at 25°C) for a glycerin-water mixture in addition to the corresponding percent volume and percent weight of glycerin. Unlike the propylene glycol properties, the values for percent volume and percent weight vary significantly, because the density of glycerin is approximately 26% higher than the density of water.

As illustrated by the v-shaped curve in Fig. 1, below, glycerin-water solutions reach their maximum freeze protection at a concentration of approximately 60% glycerin by volume. A glycerin-water solution that contains more than approximately 60% by volume glycerin will provide the same freeze protection as a less concentrated mixture. From a freeze protection standpoint, there is no reason to use a glycerin-water solution that contains more than 60% glycerin by volume.

Table 3 Glycerin Properties [10] (Portions of data are calculated or interpolated)

Glycerin Properties			
%Vol.	%Wt.*	Freezing Point (°F)	Specific Gravity at 25 °C
0%	0%	32	1.000
5%	6%	31	1.014
10%	12%	28	1.029
15%	18%	25	1.043
20%	24%	20	1.059
25%	29%	16	1.071
30%	35%	10	1.087
35%	40%	4	1.100
40%	45%	−2	1.114
45%	51%	−11	1.130
50%	55%	−19	1.141
55%	60%	−31	1.155
60%	65%	−46	1.168
65%	69%	−40	1.179
70%	74%	−25	1.193
75%	79%	−8	1.207
80%	83%	6	1.217
85%	87%	19	1.228
90%	92%	36	1.241
95%	96%	49	1.252
100%	100%	63	1.262

*% Vol. to % wt. conversion is at 25 °C

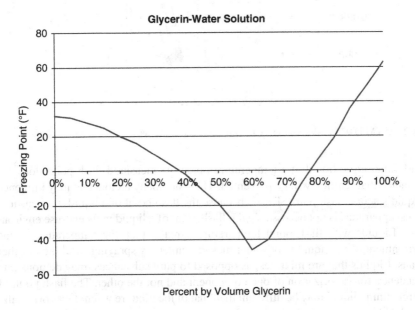

Fig. 1 Freezing point of a glycerin-water solution based on percent volume of glycerin

B   Flammability of Liquids

Liquids have many quantifiable flammability properties that vary depending on the type of liquid and the surrounding environment. The flash point is the temperature at which a liquid must be raised in order to produce sufficient vapors for flash ignition under specified test conditions. The flash point can be measured by one of many standardized test apparatus. Figure 2, below, illustrates an example of a closed-cup tester (ASTM D 93). The tester utilizes a heated stirrer (intended to maintain temperature uniformity) inserted into the test liquid. The test liquid is heated at a rate of approximately 41°F to 43°F per minute. The tester is capable of measuring the flash point of liquids between 174°F and 750°F. [11]

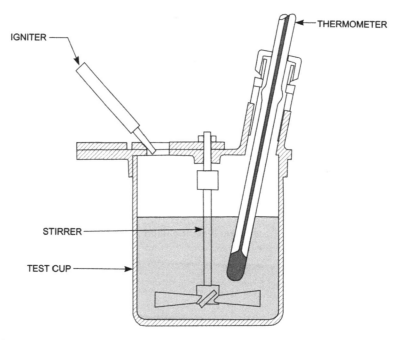

Fig. 2 ASTM D 93 Pensky-Martens Closed-Cup (PMCC) Tester

Maintaining a liquid at a temperature below its measured flash point does not guarantee that ignition will be prevented. There are many factors that may influence a liquid's actual flash point. This is because the flash point of a liquid, as measured by test apparatus, is not necessarily the flash point of a liquid in its end-use environment. Liquids with flash point temperatures greater than the temperature of the environment of the liquid may sometimes be ignited by spraying, wicking or other means. Liquids that are mixtures, as opposed to pure substances, may demonstrate a tendency for vaporization of one component and not the other. The flash point of the remaining liquid may be different than that of the mixture when it was originally tested. [11]

At some temperature above a liquid's flash point temperature, a liquid's vapor can ignite without the presence of an ignition source. This is known as the autoignition temperature (AIT). There is no known relation between a liquid's flash point and its AIT. The AIT is primarily determined by a liquid's reactivity (rate of oxidation) while the flash point is determined by a liquid's volatility (rate of evaporation). Many factors may affect a liquid's AIT. Some known factors are the concentration of the vapor given off by the liquid, the shape and size of the container, the rate and duration of heating, and the test method. [11]

Figure 3, below, is an example of a test method used to measure the AIT of liquids (ASTM E 659). In this method, the testing vessel is a glass flask surrounded by an electrically heated oven equipped with several thermocouples. During a test, a 0.1 mL sample of liquid is injected into the glass flask and is heated to a constant temperature while being observed for indications of ignition. Once the AIT is observed, both larger and smaller amounts of liquid are analyzed to determine the overall lowest AIT. [11]

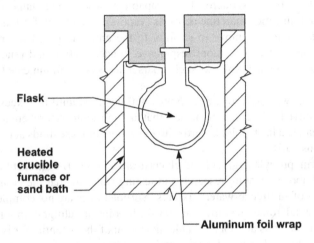

Fig. 3 ASTM E 659 Autoignition Test (Setchkin Flask Test)

Flammable substances may also have an upper flammability limit (UFL) and lower flammability limit (LFL). The UFL is the highest concentration (or lowest in the case of LFL) of gas or vapor of the liquid in air capable of producing a flash fire in the presence of an ignition source.

The following table summarizes flammability properties of chemicals permitted for use in antifreeze solutions by NFPA 13.

Table 4 Flammability Properties of Pure Antifreeze Chemicals Permitted by NFPA 13 [12]

Chemical	Flammable Limits in Air (% by volume) Lower/Upper	Flash Point (°F)	Autoignition Temperature (°F)	Boiling Point (°F)
Propylene Glycol	2.6 / 12.5	210	700	370
Glycerin	Not Provided / Not Provided	390	698	340
Diethylene Glycol	Not Provided / Not Provided	255	435	472
Ethylene Glycol	3.2 / Not Provided	232	748	387

A suspension of finely divided droplets of flammable liquid in air can yield a flammable mixture that has many of the characteristics of a flammable gas/air mixture. These droplets have the potential to burn or explode. Researchers have observed that a 10 μm diameter droplet of flammable liquid behaves like a vapor with respect to burning velocity and LFL. Droplets with diameters larger than 40 μm behave differently. [13]

Flame propagation can occur at average concentrations well below the LFL. A flammable mixture can also form at temperatures below the flash point of a combustible liquid when atomized into air. Testing shows that with fine mists and sprays (particles less than 10 μm) the combustible concentration at the lower limit is about the same as that in uniform vapor-air mixtures. However, as the droplet diameter increases, the lower limit appears to decrease. It was observed that coarse droplets tend to fall towards the flame front in an upward propagating flame, and as a result the concentration at the flame front actually approached the value found in lower limit mixtures of fine droplets and vapors. [14]

Mists made up of coarser aerosols are capable of sustaining a flame at considerably lower fuel-air ratios than fine aerosols (vapors). The reason for this lies in the ability of the droplets to move in relation to the ambient air. Mists made up of coarser aerosols prove to be more responsive to acceleration and random movement than that of finer aerosols. As such, coarser aerosols communicate flame more readily. [13]

In the case of water-glycols, flash points will not exist until the excessive water has been removed. Research indicates that a high-temperature environment is required to realize a flash point hazard with the vapors of these fluids at normal pressure conditions. [15]

In pure form, propylene glycol and glycerin are Class IIIB Combustible Liquids. As discussed above, existing research and testing suggests that the combustibility characteristics of antifreeze-water mixtures in droplet form are not completely characterized by standardized test methods for flash point or autoignition temperature. As such, these methods are not a reliable indication of the potential for ignition of a liquid dispersed into droplets. Under certain conditions, atomized antifreeze-water mixtures can combust when sprayed onto an ignition source. Increasing the concentration of the antifreeze in the antifreeze-water solution increases the combustibility of the solution.

Antifreeze solutions of propylene glycol and water have been permitted in sprinkler systems for more than 50 years at concentrations as high as 60% by volume, which is equal to 60% by weight. However, the following disclaimer is included in the MSDS for a premix antifreeze solution specifically intended for sprinkler systems:

Fire and Explosion Hazards – Heat from fire can generate flammable vapor. When mixed with air and exposed to ignition source, vapors can burn in open or explode if confined. Vapors may travel long distances along the ground before igniting and flashing back to vapor source. Fine sprays/mists may be combustible at temperatures below normal flash point. Aqueous solutions containing less than 95% propylene glycol by weight have no flash point as obtained by standard test methods. However aqueous solutions of propylene glycol greater than 22% by weight, if heated sufficiently, will produce flammable vapors. Always drain and flush systems containing propylene glycol with water before welding or other maintenance. [16]

The disclaimer above identifies the potential for vapors of aqueous solutions that contain certain concentrations of propylene glycol to combust. It is important to consider this potential for combustion when dealing with aqueous solutions that contain flammable liquids (e.g. propylene glycol and glycerin). Furthermore, the disclaimer identifies that fine sprays/mists may be combustible at temperatures below their normal flash point.

The discussion above describes the complexity of whether a certain antifreeze solution has the potential to ignite when supplied through automatic sprinkler systems. Existing research indicates that under certain conditions the energy released during a fire condition could increase upon interaction with certain antifreeze-water mixtures currently permitted by NFPA 13, 13D and 13R. [17] [18] Recent testing conducted by UL [4] demonstrates that, under certain conditions, a large-scale sustained ignition is possible from the discharge of certain sprinkler systems containing antifreeze solutions. The intent of the Phase II testing is to more completely investigate the potential for large-scale ignition of flash fires from antifreeze solutions and to investigate the impact on a sprinkler system's ability to control a fire condition and maintain tenable conditions.

C   Sprinkler Droplet Sizes and Distributions

Droplet sizes and distributions produced by automatic sprinklers have been studied using a variety of techniques. Measurements of the droplet sizes produced by automatic sprinklers are relatively complex because the droplet size distribution measured is expected to vary with several factors including:

- Position with respect to the sprinkler in three dimensions
- Sprinkler model
- Operating pressure/flow rate
- Liquid supplied to the sprinkler, e.g. water or antifreeze solution
- Surrounding air currents, including fire induced flows

Even with all of the variables above held constant, measurements include a range of droplet sizes and not a single uniform droplet size. Additionally, it is possible for sprinklers operating with identical k-factors and pressures to have different spray patterns. Sprinklers that have identical orifice sizes (k-factor) can have varying geometric parameters such as arms, deflectors or tines. Changes in any of these geometric parameters may substantially alter the droplet size and distribution. For example, the figure below illustrates sprinkler discharge from two sprinklers with the same k-factor operating under the same pressure, but with spray distribution patterns that are significantly different.

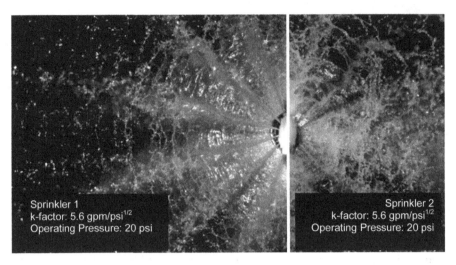

Fig. 4 Spray distribution from automatic sprinklers. (Courtesy: Prof. André Marshall, University of Maryland)

Many of the existing methods are point measurement techniques that only measure data at a single point. Point measurement techniques are capable of measuring droplet size and velocity and work well for spherical droplets. However, sprinkler droplets are not always spherical.[19] In addition, point measurements must be taken at various locations in the sprinkler flow so that the results are temporally and spatially averaged. This limits measurement accuracy because fire sprinkler sprays are unsymmetrical and unsteady. Certain areas of the spray distribution are denser than others which may cause results to vary based on measurement locations. [19]

Studies of standard orifice, pendent spray fire sprinklers indicate droplet sizes between approximately 200 and 3,000 µm. [19] This approximation agreed with existing research which indicated that droplets larger than approximately 5,500 µm in diameter are unstable and break up into smaller droplets, predominantly in the range of 1,000 to 2,000 µm. [20] Previous research indicates that while a large number of very small drops are present, they comprise a small portion of the total water volume. Data indicates that 98% of the water from standard orifice fire sprinklers is contained in droplets larger than 200 µm in diameter. [19] A study of residential sprinklers measured water droplets ranging from an arithmetic mean of 200 to over 500 µm, depending on location. [21] However, droplets with diameters of less than 100 µm were measured. [21]

D   Phase I Testing

During Phase I of this project a series of preliminary tests were sponsored and conducted by Underwriters Laboratories. Tests were conducted in UL's large-scale test facility in Northbrook, IL and several of the tests were witnessed by CCI on behalf of the Fire Protection Research Foundation.

Initial tests were conducted with a small ceiling above an elevated pan of heptane using residential pendent sprinklers with nominal k-factors of 3.1 and 4.9 gpm/psi$^{1/2}$. The tests used premixed solutions of 70% glycerin and 60% propylene glycol with water. The tests indicated the potential for large-scale ignition of a 70% glycerin solution using a 3.1 k-factor sprinkler at an operating pressure of 100 psi. This large-scale ignition resulted in flames surrounding the majority of the sprinkler spray. A similar large-scale ignition did not occur for initial tests with 60% propylene glycol solutions or tests using a 4.9 k-factor sprinkler at an operating pressure of 50 psi.

Further tests were conducted in a three sided room measuring approximately 12 feet by 12 feet with a ceiling height of 8 feet. A single sprinkler with a k-factor of 3.1 was located in the center of the ceiling for each test. The majority of the room tests were conducted using a nominal 12-inch cast-iron pan with cooking oil as the initial fire source. An electric cooktop was used to heat the pan and ignite the cooking oil. One room test was conducted with a pan of heptane as the initial fire instead of the cooking oil. In various tests, the sprinkler was supplied with water only as well as premixed solutions of 70% glycerin, 50% glycerin, and 60% propylene glycol in water. Sprinkler operating pressures of 20, 100, and 150 psi were investigated.

Test results in the room configuration ranged from extinguishment of the fire to large-scale, sustained ignition of the antifreeze solution. Preliminary observations during the tests indicated that the results depend, at a minimum, on a combination of the following factors:

- Location of the initial fire with respect to the sprinkler
- Initial fire source
- Type of sprinkler and operating pressure
- Type and concentration of antifreeze solution

Large-scale, sustained ignition of the 70% glycerin solution supplied at 100 psi occurred when the initial fire was in close proximity to the sprinkler, but the initial fire was controlled using the same concentration of antifreeze at the same operating pressure when the initial fire was located farther from the sprinkler. Large-scale ignition of the 60% propylene glycol solution occurred in the room configuration during a cooking oil fire, but did not occur in the open configuration during a heptane fire. Large-scale ignition of the antifreeze solution did not occur in any of the tests with the 50% glycerin solution.

Preliminary observations during the UL testing indicate the following:

- Large-scale ignition of antifreeze solutions occurred in certain tests for 70% solutions of glycerin and 60% solutions of propylene glycol with water.
- Large-scale ignition of antifreeze solutions of 50% glycerin with water did not occur for any of the tested configurations.

Preliminary observations from the tests highlighted the need for further research into the effectiveness of currently permitted antifreeze solutions and consideration of their suitability for use in sprinkler systems.

III
Phase II Test Plan and Setup

The Phase II testing was intended to further study the potential for contribution of antifreeze solutions to fire conditions. The Phase II test plan was separated into two scopes. Scope A tested antifreeze solutions for the potential to create a large-scale ignition of the spray when discharged through sprinklers onto a fire. Scope B tested antifreeze solutions for their impact on a sprinkler system's ability to control a fire condition and maintain tenable conditions.

Tests were conducted with premixed solutions of propylene glycol and glycerin with water obtained from a single commercial distributor. Application of the test results is limited to the solutions tested and not to other formulations of antifreeze solutions that were not tested. Phase II tests were conducted at UL's fire test facility in Northbrook, IL and a Summary of Fire Test Data was provided by UL in Reference [22].

A Scope A: Fire Tests for Spray Ignition Using Sprinklers

Scope A was developed to investigate the potential for ignition of antifreeze solutions supplied by automatic sprinklers. The tests were designed to use a strong, continuous ignition source to identify whether flammable mixtures of antifreeze were created by the antifreeze spray. The tests used several models of residential sprinklers to investigate their impact of the potential for ignition.

Scope A tests were conducted without an enclosure, other than the walls and roof bounding the laboratory. As discussed in the report from Phase I of this project [3], the difference between a flash fire and an explosion is the degree of confinement of the flash fire. Because an explosion could not occur in this context without a flash fire, the flash fires themselves were used as criteria for the tests without the need to evaluate a resulting, enclosure dependant, explosion. While the test setup was designed to avoid explosions within the laboratory, the confinement of flash fires can produce over-pressurizations or explosions.

Code Consultants, Inc., *Antifreeze Solutions in Home Fire Sprinkler Systems*,
SpringerBriefs in Fire, DOI 10.1007/978-1-4614-3840-3_3,
© Fire Protection Research Foundation 2010

The test setup for Scope A is illustrated in Fig. 5, below.

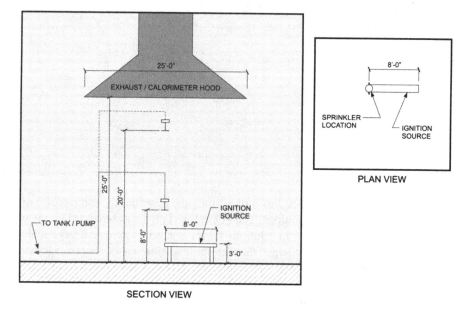

Fig. 5 Scope A test setup

The test setup included a long ignition source that was designed to extend radially from the sprinkler location. The long ignition source allowed a single test to investigate the potential for ignition over a range of locations within the spray pattern. The arrangement allowed for multiple sprinkler heights to be tested and data was collected to allow for heat release rate measurements using oxygen consumption calorimetry.

Initial testing was conducted to investigate appropriate ignition sources. Ignition sources investigated included:

- 6" wide and 12" wide rectangular pans of heptane extended radially from the point directly below the sprinkler.
- 4-nozzle heptane spray burners under a metal grate (the metal grate functions as a hot surface to vaporize antifreeze solution).
- Electric range heating elements (also functioning as a hot surface to vaporize antifreeze solutions).

Figure 6, below, illustrates each of the ignition sources investigated as part of Scope A.

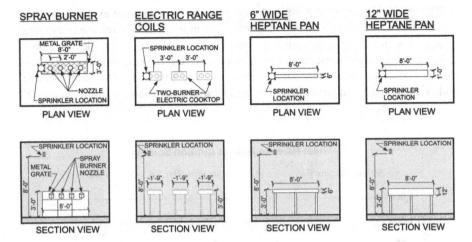

Fig. 6 Scope A ignition sources

The ignition sources are also shown in the photographs below.

Fig. 7 Ignition source photographs

The heat release rate of each of the ignition sources, with the exception of the electric range coils, is illustrated in the following graph. The heat release rates were measured using an oxygen consumption calorimeter. Because the electric range coils are heated by electricity and not combustion, the oxygen consumption calorimeter could not measure the heat release rate during that test. The electric range coils were tested based on the high temperature of the coils and not due to their total heat release rate.

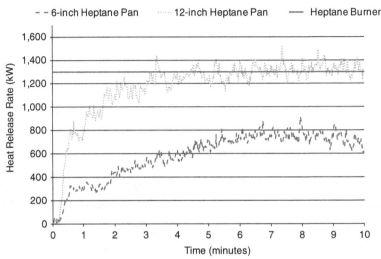

Fig. 8 Comparison of ignition source heat release rates

The heat release rate of the pan fires increases for several minutes after ignition before reaching a steady heat release rate. For tests with the pan fires sprinkler flow was initiated three minutes after ignition of the pan fire to allow the pan fire to reach a nearly steady heat release rate.

The following graph shows the maximum heat release rate of various household furniture items [23] in comparison with the heat release rate of the heptane spray burner.

Figure 9, below, shows that the heat release rate of the ignition sources is less than the maximum heat release rates of some common household furniture items. The data above indicates heat release rates measured under the specific conditions tested without the benefit of sprinkler protection. Thus, in a residence protected by automatic sprinklers, the heat release rate at the time of sprinkler activation could be less than the heat release rates illustrated in Fig. 9.

An estimate of the heat release rate at the time of sprinkler activation was calculated using the Sprinkler/Detector Response routine in the computer fire model FPETool.

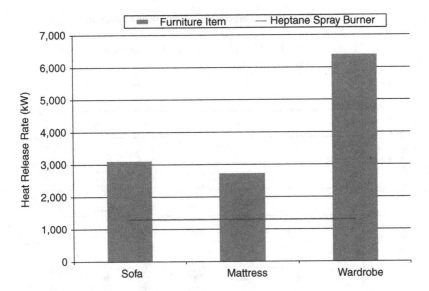

Fig. 9 Heat Release Rate of Home Furnishings

[24] The model was originally developed by the National Bureau of Standards, now the National Institute of Standards and Technology, as DETACT-QS. The DETACT-QS model is a basic computer fire model that calculates the temperature and velocity at a sprinkler based on correlations developed by Alpert [25] and combines them with a lumped-capacitance heat transfer model to estimate the time and heat release when sprinkler activation is calculated to occur. The model is designed to large, open spaces and does not account for the effect of the room enclosure on the temperatures at the sprinkler. Thus, for residential scale rooms the model typically over-predicts the fire size at the time of sprinkler activation. The results below are based on the following parameters:

Input	Value
Ambient temperature	70 °F
Sprinkler activation temperature	155 °F
Response Time Index (RTI)	91 $(ft.s)^{1/2}$
Horizontal Distance from Fire to Sprinkler	10.6 ft
Fire growth rate	Medium t-squared (growth time of 300 s to reach 1,055 kw)

The following graph shows the calculated heat release rate at the time of sprinkler activation along with the heat release rate of the heptane spray burner fire used for the Scope A tests.

Calculated Heat Release Rate at Sprinkler Activation

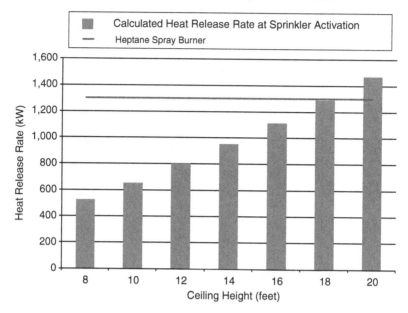

Fig. 10   Heat release rate at sprinkler activation based on ceiling height

The results summarized in Fig. 10, above, show that the heat release rate of the ignition source used in the Scope A tests was generally conservative for spaces with ceiling heights of less than 20 feet. The calculated heat release rate at the time of sprinkler activation for ceiling heights of less than 20 feet are less than the heat release rate of heptane burner ignition source used in the Scope A tests. The calculated heat release rate for a ceiling height of 20 feet was within approximately 5% of the heat release rate of the heptane burner used in the Scope A tests.

Ignition sources were tested using solutions of 50% propylene glycol and 60% propylene glycol supplied from a residential pendent sprinkler with a k-factor of 3.1. Prior testing indicated that a 60% propylene glycol solution can be ignited by a kitchen grease fire when supplied from a k3.1 sprinkler. Thus, the ignition source selected should be capable of igniting the 60% propylene glycol solution supplied through a k3.1 sprinkler. It was unclear prior to the start of testing whether the 50% propylene glycol solution would be ignited.

The ignition sources selected for the Scope A testing are very unlikely to be extinguished during sprinkler activation. This is unlike most home fire conditions that would be expected to reduce in intensity upon the application of water. Some increase in heat release rate could be expected for the ignition sources, but observations of flash fires or ignition of the spray away from the fire source were considered an immediate failure. The initial testing was also intended to validate the use of a variable sprinkler operating pressure. Varying the sprinkler operating pressure allowed each test to collect data for a range of sprinkler operating pressures. This approach reduced the overall number of tests conducted and helped yield more complete results.

Following the initial testing, a series of tests was conducted to investigate the potential for ignition of select concentrations of antifreeze for the following variables:

Variable	Values Tested
Antifreeze concentration	• Propylene glycerol ○ 40%, 45%, 50%, 60% • Glycerin ○ 50%, 55%
Antifreeze temperature	• Ambient 80-90 °F • Elevated 140 °F
Sprinkler height	• 8 ft • 20 ft
Horizontal position of ignition source	Considered through the use of a long ignition source that extended radially from the sprinkler
Sprinkler operating pressure	10 to 150 psi (varied in 10 psi increments)
Sprinkler type and nominal k-factor	• Fixed deflector residential pendent (k3.1, k4.9, k7.4) • Drop-down deflector (concealed) residential pendent (k4.9, k5.8) • Residential sidewall (k4.2, k5.5)

The majority of the testing was conducted with solutions of 40%, 50%, and 60% propylene glycol as well as 50% glycerin. Select tests of 45% propylene glycol and 55% glycerin were used to evaluate the sensitivity of the results to the antifreeze concentration.

Ceiling heights of 8 ft and 20 ft were used to evaluate a range of residential applications. The 8 ft ceiling height is typical of many residential spaces and the 20 ft ceiling height is intended to account for a tall, double-height space in a residential occupancy. It was theorized prior to the initial testing that the atomization and dispersion of the droplets in the sprinkler spray would behave differently for varying ceiling heights. The initial testing confirmed that the spray distribution reaching the fire sources changes with the height of the sprinkler.

The Phase I testing demonstrated that the position of the ignition source within the sprinkler spray significantly impacted the potential for ignition of the spray. The long ignition source extending radially from below the sprinkler was used to allow a single test to generate data for a range of ignition source locations.

Data was gathered for a wide range of sprinkler operating pressures by varying the operating pressure during each test. The low pressure (10 psi) was intended to capture data near the minimum flow rates that would be permitted for the larger orifice sprinklers in the test plan. The high pressure (150 psi) was intended as a high pressure anticipated for a typical residential occupancy. In some instances the tests were conducted starting at a higher operating pressure greater than 10 psi or were terminated prior to reaching 150 psi based on the data to be collected from that test.

Due to the complex nature of the droplet size and sprinkler spray distribution produced during sprinkler discharge, several different types of sprinklers were selected for the Scope A testing. This approach was used to develop information on how changes in sprinkler geometry (deflector, arms and tines) and orifice size impacted the results.

B   Scope B: Room Fire Tests of Sprinkler Effectiveness

The Scope B tests were intended to investigate the effectiveness of residential sprinklers using an antifreeze solution compared with water alone. The Scope B tests were not intended to investigate the potential for large-scale ignition of the sprinkler spray.

The Scope B testing is similar to the UL 1626 fire test, with certain additional variables considered as outlined in the table below.

Variable	Values Tested
Antifreeze solutions	• 50% Glycerin • 40% Propylene glycol (single test) • Water alone
Ceiling height	• 8 ft

Sprinkler operating pressure/flow rate	• Minimum permitted flow based on NFPA 13D design criteria
	  ○ Pendent: 18 gpm one sprinkler / 13 gpm each for two sprinklers
	  ○ Sidewall: 24 gpm one sprinkler / 17 gpm each for two sprinklers
	• 80 psi
	• 150 psi

Sprinkler type, temperature rating, and nominal k-factor	• Ordinary temperature fixed deflector residential pendent (k3.1, k4.9) • Ordinary temperature residential sidewall (k4.2)
Fire Source	• UL 1626 fuel package • Furnished living room (sofa, chair, tables)

The tests are designed to directly compare the performance of sprinkler systems supplied with antifreeze solutions to the performance of sprinkler systems supplied with water alone. The tests measured temperature at several locations within the room to evaluate tenability in accordance with the criteria specified in UL 1626. In addition, the test setup included sprinklers installed within the room that were designed to evaluate whether the fire condition would be expected to overwhelm the sprinkler system. Figure 11, below, illustrates the test setup for Scope B.

For the purposes of the Scope B testing, a ceiling height of 8 ft was used. This ceiling height was intended to represent that of a typical residential dwelling.

Similar to the Scope A tests, Scope B tests included multiple sprinkler operating pressures, but the pressure was not varied during each test. The low flow tests were intended to match the NFPA 13D criteria of 18 gpm for the activation of the first sprinkler and 13 gpm per sprinkler for the activation of two sprinklers. For the sidewall sprinklers a minimum flow rate of 24 gpm for the first sprinkler and 17 gpm per sprinkler for the activation of two sprinklers was required based on the listing of the sprinkler. Higher sprinkler operating pressures of 80 and 150 psi were also tested to evaluate their impact on the results.

A range of sprinkler types and models were tested in Scope B. Two sprinklers were located within the test room in accordance with UL 1626 to control the fire condition and a third sprinkler was located near the doorway to the enclosure, as illustrated in Fig. 11, below, to investigate the potential for activation of sprinklers away from the area of fire origin. The test enclosure measured 32 ft by 16 ft by 8 ft high, which was within the listed spacing of the k4.9 and sidewall sprinklers. The enclosure was somewhat larger than the 14 ft by 14 ft listed spacing of the k3.1 sprinkler, so the larger enclosure provided a severe test of the antifreeze solution.

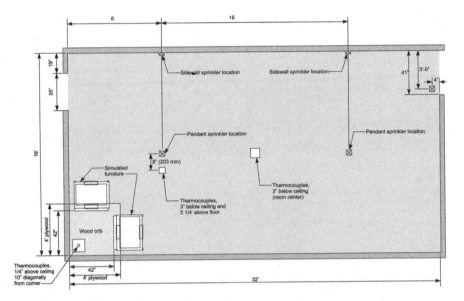

Fig. 11 Scope B test setup

The tests primarily used the fuel package specified in UL 1626 that consists of a wood crib ignited by a pan of heptane that is positioned adjacent to two simulated furniture ends. The potential for fire spread is evaluated by locating the fuel package in the corner of the room with walls covered with wood paneling. In addition to tests with the UL 1626 fuel package, a test was also conducted with a fuel package typical of a residential living room. The fuel package consisted of a sofa, chair, end table, and coffee table, along with a trash can filled with paper.

Failure criterion for the Scope B testing was based on the UL 1626 fire control criteria. Based on these criteria, residential sprinklers installed in a fire test enclosure with an 8-ft ceiling are required to control a fire for 10 minutes with the following limits:

1. The maximum gas or air temperature adjacent to the sprinkler 3 inches below the ceiling at two locations within the room must not exceed 600 °F.
2. The maximum temperature 5 feet 3 inches above the floor at a specified location within the room must be less than 200 °F during the entire test. This temperature must not exceed 130 °F for more than a 2 minute period.
3. The maximum temperature ¼ inch behind the finished surface of the ceiling material directly above the test fire must not exceed 500 °F.
4. No more than two residential sprinklers in the test enclosure can operate.

Any variation from the limits outlined above was considered an immediate failure. [26]

IV
Phase II Test Results

A Scope A – Spray Ignition

Initial tests were conducted to investigate potential ignition sources. The tests used solutions of 50% and 60% propylene glycol to investigate the effectiveness of each ignition source in igniting antifreeze sprays. The following graph compares the increase in heat release rate due to ignition of 60% propylene glycol antifreeze spray for each of the ignition sources.

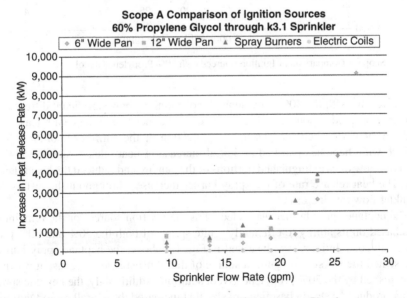

Fig. 12 Scope A Comparison of Ignition Sources with 60% Propylene Glycol

Code Consultants, Inc., *Antifreeze Solutions in Home Fire Sprinkler Systems,*
SpringerBriefs in Fire, DOI 10.1007/978-1-4614-3840-3_4,
© Fire Protection Research Foundation 2010

Each of the ignition sources, with the exception of the electric range coils, was able to ignite the 60% propylene glycol solution. The increase in heat release rate from the spray burner assembly was somewhat higher than the other ignition sources at the same sprinkler flow rate. Note that the pan and spray burner tests were terminated early due to the size of the resulting fire condition.

Figure 13, below, shows the increase in heat release rate as a function of sprinkler flow rate for a 50% propylene glycol solution using each of the ignition sources that ignited the 60% solution.

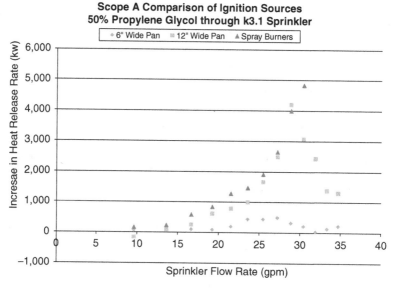

Fig. 13   Scope A Comparison of Ignition Sources with 50% Propylene Glycol

The results with the 50% propylene glycol solution show significant differences between the ignition sources. There was very little increase in the heat release rate of the 6-inch wide heptane pan upon application of the antifreeze solution. The 12-inch wide heptanes pan had an initial increase in heat release rate, but higher sprinkler flow rates extinguished portions of the pan fire and reduced the heat release rate. The heat release rate of the spray burner increased throughout the test as the sprinkler flow rate increased.

The heptane spray burner was selected as the ignition source for the remaining tests based on its ability to efficiently ignite sprays of both the 50% and 60% propylene glycol solutions. As illustrated in Fig. 13, above, the heptane spray burner represented the worst-case ignition source of those investigated, because it was not extinguished by the 50% propylene glycol solution. Additionally, the heptane spray burner produced a steady baseline fire size that increased the overall reproducibility and reliability of the ignition source.

Tests were conducted by lighting the heptane burners, adjusting the heptane flow rate, allowing for 2 minutes of heating, and flowing antifreeze solution to an open sprinkler. The sprinkler operating pressure was typically varied during each test from 10 psi to 150 psi, unless the test was terminated early due to the growth of the fire condition.

The tests investigated the impact of several variables in causing ignition of anti-freeze sprays.

1   Sprinkler

Tests of 50% propylene glycol solution were conducted for the full range of sprinklers investigated. The graph that follows shows the increase in heat release rate due to an antifreeze spray of 50% propylene glycol for the range of sprinklers.

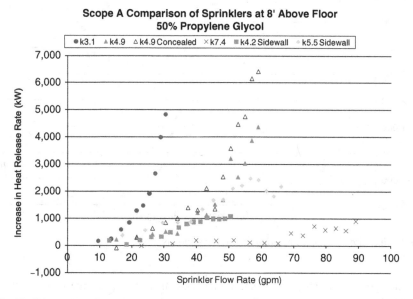

Fig. 14 Comparison of Sprinklers at 8' Above Floor with 50% Propylene Glycol

The results presented in Fig. 14, above, show that a 50% propylene glycol solution results in a significant increase in the size of the initial fire when supplied by certain sprinklers. Data for three of the six sprinklers tested shows an increase of more than 4,000 kW or 300% in the heat release rate due to the application of 50% propylene glycol antifreeze solution depending on the operating pressure. Very little ignition of the spray was observed during the test with the k7.4 pendent sprinkler. The results for the two k4.9 pendent sprinklers show that the portion of the spray that is ignited can differ for sprinklers with the same k-factor. For example, at a flow of approximately 55 gpm the increase in heat release rate during the test with the k4.9 pendent sprinkler was approximately 3,000 kW compared with more than 4,500 kW in the test with a concealed sprinkler. Further testing primarily used the k3.1 and k4.9 concealed sprinklers based on the results outlined above.

2 Antifreeze Solution

Scope A tests were conducted for solutions of 40%, 45%, and 50% propylene glycol as well as 50% and 55% glycerin. Results of the tests are summarized in Fig. 15, below, which shows the increase in heat release rate due to the application of each antifreeze solution using the same sprinkler and ignition source.

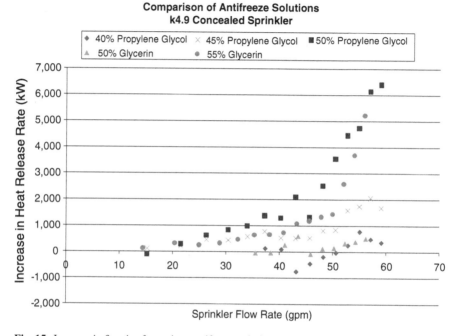

Fig. 15 Increase in fire size for various antifreeze solutions

The results presented above show increases in heat release rate of more than 6,000 kW or 500% for the 50% propylene glycol and 55% glycerin solutions at certain flow rates. This is due in large part to ignition of the antifreeze spray extending away from the initial fire condition. A significantly lower increase in heat release rate was measured for the 45% propylene glycol solution, which showed little ignition of the sprinkler spray away from the ignition source. The application of anti-freeze solutions of 40% propylene glycol and 50% glycerin resulted in much smaller changes in heat release rate during otherwise identical test conditions. The 40% propylene glycol and 50% glycerin solutions resulted in very similar changes in the heat release rate of the fire condition. Although there was some increase in the heat release rate that was measured for both solutions at certain operating pressures, flames were not observed to extend away from the initial fire source.

Figure 16 and Figure 17, below, illustrate the maximum increase in heat release rate caused by 50% glycerin solution for tests with sprinklers at 8 ft and 20 ft above the floor, respectively. The maximum heat release rate measured for the test at 8 ft was approximately 3,300 kW and 2,800 kW for a test at 20 ft, compared with a baseline ignition source heat release rate of approximately 1,400 kW.

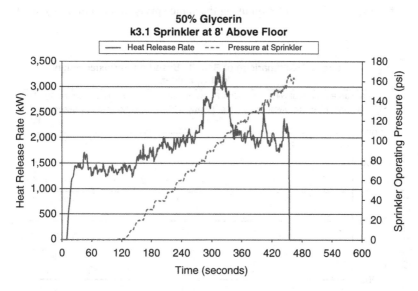

Fig. 16 Detailed results for 50% glycerin supplied through k3.1 sprinkler at 8 ft

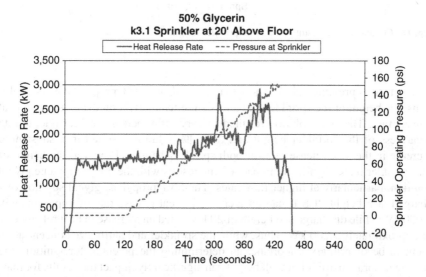

Fig. 17 Detailed results for 50% glycerin supplied through k3.1 sprinkler at 20 ft

3   *Sprinkler Height*

Tests were conducted for solutions of 40%, 50% and 60% propylene glycol for sprinkler heights of 8 feet and 20 feet. Results of the tests are summarized in Fig. 18, below, which shows the increase in heat release rate due to the change in ceiling height for each antifreeze solution using the same sprinkler and ignition source.

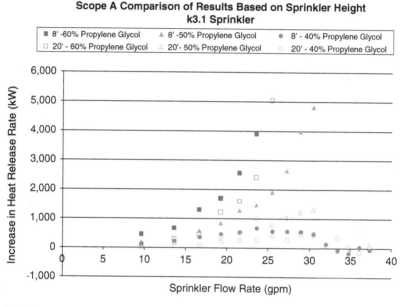

Fig. 18 Comparison of Results Based on Sprinkler Height

The results presented above show that for 40% and 60% propylene glycol solutions, the height of the sprinkler had a less significant effect on the increase in heat release rate. The 40% solution resulted in very little increase in heat release rate regardless of the sprinkler height and the 60% solution resulted in a substantial increase in the heat release rate for both sprinkler heights. However, the height of the sprinkler had a significant impact on the results with the 50% propylene glycol solution, particularly at higher flow rates. The 50% propylene glycol solution discharged at a height of 8 ft had an increase in heat release rate of approximately 5,000 kW while discharge at a height of 20 ft yielded an increase in heat release rate of approximately 1,200 kW. Thus, while the sprinkler and antifreeze concentration seem to be of primary importance in determining the potential for ignition, the change in spray distribution with height can significantly impact the results for marginal solutions.

4 Temperature of Antifreeze Solution

Tests were conducted that compared the performance of glycerin solution at ambient temperature (80°F to 90°F) and glycerin solution heated to 140°F. Results of the tests are summarized in Fig. 19, below, which illustrates the increase in heat release rate for heated and unheated 50% glycerin solutions.

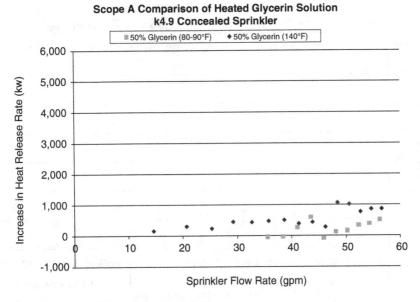

Fig. 19 Comparison of Heated Glycerin Solution

The results presented above shows a minor difference in heat release rate during tests with ambient temperature and heated glycerin solutions. Each of the solutions produced a maximum increase in heat release rate of approximately 500 to 1,000 kW. While there may be some difference based on temperature over the range investigated, it appears that the effect of temperature is minor compared with the impact of solution concentration.

B Scope B – Room Fire Tests

Failure criterion for the Scope B testing was based on the UL 1626 fire control criteria. Based on these criteria, residential sprinklers were installed in a fire test enclosure with an 8-ft ceiling and are required to control a fire for 10 minutes within the limits established by the UL 1626 fire control criteria. UL 1626 includes provisions for extending the duration of the test to 30 minutes if continued burning is observed at 10 minutes, but the test duration was limited to 10 minutes for the purposes of this comparison.

1   Temperature 3 inches Below Ceiling

Tests were conducted to ensure that the maximum temperature adjacent to the sprinkler 3 inches below the ceiling did not exceed 600°F. Figure 20, below, illustrates the results of these tests.

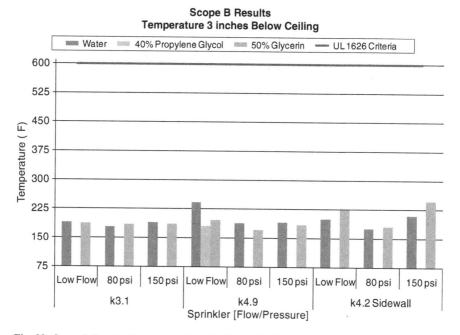

Fig. 20 Scope B Results Temperature 3 inches Below Ceiling

The results in Fig. 20, above, show that water, 40% propylene glycol, and 50% glycerin demonstrate similar performance. Regardless of variation in sprinkler operating pressure and k-factor, both of the antifreeze solutions and water did not exceed a measured temperature of 246°F. This is well below the maximum temperature of 600°F specified in the UL 1626 fire control criteria.

2   Temperature at 5'-3" Above Floor

Temperature results at 5'-3" above the floor are illustrated in Fig. 21 and Fig. 22, below. Figure 21 shows the maximum temperature measured during the test, which is limited by UL 1626 to 200°F, and Fig. 22 shows the temperature that is sustained for 2 minutes during the test, which must be less than 130°F based on the criteria in UL 1626.

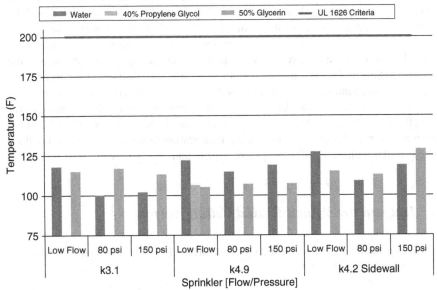

Fig. 21 Maximum Temperature 5'-3" Above Floor

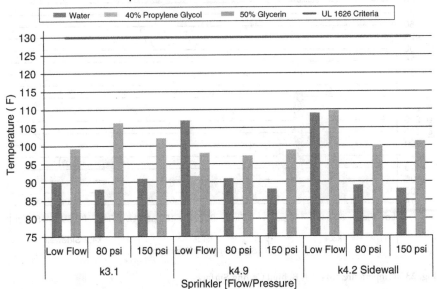

Fig. 22 2 minute Sustained Temperature at 5'-3" Above Floor

All of the tests remained well below the temperature criteria specified in UL 1626. The maximum temperature for water and 50% glycerin were each slightly higher than 125°F compared with a criteria of 200°F. For the low flow condition and the 2 minute temperature criteria, the results with the 50% glycerin solution were better than water for the test with the k4.9 sprinkler, the results with water were better for the k3.1 sprinkler, and the results with the k4.2 sidewall sprinkler were nearly the same. The results for the 2 minute temperature criteria in the tests at 80 psi and 150 psi show somewhat higher temperatures with the 50% glycerin solution compared with water. This may be due in part to the flow rate of glycerin solution being lower than the flow rate of water at the same pressure, which should be accounted for in the design of a sprinkler system. Overall, the temperature results at 5'-3" above the floor were similar with water, 40% propylene glycol, and 50% glycerin.

3   Temperature ¼-inch Behind Ceiling Surface

The temperature results at ¼-inch behind the ceiling surface above the fire are illustrated in Fig. 23, below.

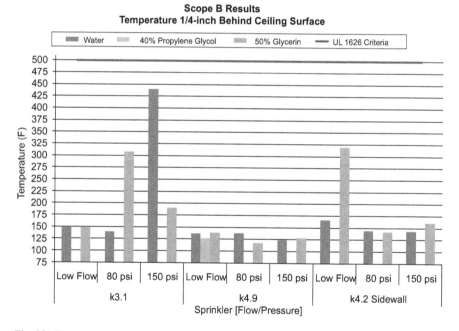

Fig. 23 Temperature 1/4–inch Behind Ceiling Surface

As shown in the figure above, the majority of the tests had similar results and all of the tests remained within the criteria specified by UL 1626. In two of the configurations the test with 50% glycerin solution had significantly higher temperatures than the similar test with water and in one of the configurations the test with water had significantly higher temperatures than the similar test with glycerin solution. The highest measured temperature behind the ceiling material was during the test with the k3.1 sprinkler supplied with water at 150 psi. This result is likely due to the test room being larger than the listed protection area of the sprinkler; however, the same test configuration with 50% glycerin solution better controlled the fire condition.

4   Number of Sprinkler Activated

The UL 1626 criteria allows no more than two of the three sprinklers in the room to activate for a successful test. Figure 24, below, shows that the number of sprinklers activated met this criteria for each of the tests.

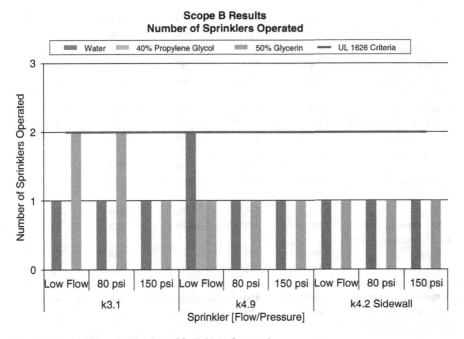

Fig. 24   Scope B Results Number of Sprinklers Operated

Two sprinklers were activated in the enclosure during two of the tests with glycerin solution and one of the tests with water. For the remaining tests only a single sprinkler activated. Based on the results of these tests, as illustrated above, the UL 1626 criteria was satisfied.

5 Scope B Summary

The results for Scope B are summarized in Table 5, below, along with the UL 1626 criteria.

Table 5 Scope B Test Results

Sprinkler [Flow/Pressure] Solution	Temperature 3" Below Ceiling Maximum (°F)	Temperature 5'-3" Above Floor Maximum (°F)	2-Minute (°F)	Temperature Behind Ceiling Material Maximum (°F)	No. of Sprinklers Activated Maximum
UL 1626 Criteria	600	200	130	500	2
k3.1 Low Flow					
Water	190	118	90	148	1
50% Glycerin	188	115	99	148	2
k3.1 80 psi					
Water	178	100	88	140	1
50% Glycerin	184	117	106	308	2
k3.1 150 psi					
Water	190	102	91	440	1
50% Glycerin	186	113	102	190	1
k4.9 Low Flow					
Water	241	122	107	137	2
40% Propylene Glycol	180	106	92	127	1
50% Glycerin	196	105	98	139	1
k4.9 80 psi					
Water	189	115	91	138	1
50% Glycerin	172	107	97	117	1
k4.9 150 psi					
Water	191	119	88	124	1
50% Glycerin	185	107	99	128	1
k4.2 Sidewall Low Flow					
Water	200	127	109	166	1
50% Glycerin	223	115	110	319	1
k4.2 Sidewall 80 psi					
Water	175	109	89	144	1
50% Glycerin	180	113	100	142	1
k4.2 Sidewall 150 psi					
Water	209	119	88	143	1
50% Glycerin	246	129	101	161	1
Furniture Fire k4.9 Low Flow					
50% Glycerin	165	96	94	104	1
Without Sprinklers	**>1,074**	**>545**	**>130**	**>571**	**N/A**

In addition to the results of tests with the UL 1626 fuel package, Table 5 also includes the results of a test conducted with living room furniture. The test used 50% glycerin solution supplied to a k4.9 sprinkler at 18 gpm. The fire was controlled by one sprinkler. The results of the test indicate that the UL 1626 fuel package is a more severe test of the sprinkler system than the living room furniture fuel package. The temperatures measured during the test with actual furniture were lower than any of the tests with the UL 1626 fuel package.

Table 5 also includes results of a UL 1626 type test conducted by Underwriters Laboratories without the use of sprinklers. The test without sprinklers was conducted as part of a prior research project and used a 12 ft by 24 ft enclosure meeting the requirements of UL 1626. The test was terminated after less than 4 minutes when the temperature in the room exceeded 1,000°F. While all of the Scope B tests with antifreeze solutions and water maintained temperatures within the UL 1626 criteria for a full 10 minutes, a similar test without sprinklers resulted in flashover of the enclosure in less than 4 minutes. The results demonstrate the effectiveness of water as well as solutions of 40% propylene glycol and 50% glycerin in controlling home fire conditions represented by UL 1626.

V
Analysis

The fire test program conducted as part of Phase II of this project was intended as an empirical evaluation of the potential for ignition and impact on sprinkler system effectiveness of various antifreeze solutions. The following observations and analysis provides additional insight into the fire test data discussed above.

A Observations

The Scope A ignition tests conducted with the heptane spray burner were unique in that the ignition source had a steady heat release rate that was not significantly impacted by the antifreeze spray and the ignition source could not readily be extinguished by the antifreeze spray. This allowed the contribution of the antifreeze solution to the fire condition to be accurately measured, since the heat release rate of the ignition source itself could not be enhanced by the antifreeze spray, i.e. the fuel contributed by the spray burner was a function of the heptane flow rate and was not impacted by the antifreeze spray. In addition, it provided a conservative assessment of the potential for ignition, because many fire sources would be extinguished even by antifreeze solutions that had a significant increase in heat release rate during the tests.

Even the lowest antifreeze concentrations tested resulted in some increase in heat release rate when exposed to the heptane spray burner fire for certain test conditions. For example, Figure 15 shows that the 40% propylene glycol solutions result in some increase in the heat release rate of the initial fire at certain flow rates. The increase in heat release rate can be included in one of two categories:

1. Ignition of antifreeze spray that reaches the ignition source; or
2. Ignition of antifreeze spray extending away from the ignition source.

Code Consultants, Inc., *Antifreeze Solutions in Home Fire Sprinkler Systems*,
SpringerBriefs in Fire, DOI 10.1007/978-1-4614-3840-3_5,
© Fire Protection Research Foundation 2010

Based on the tests conducted, it appears that some portion of the antifreeze spray reaching the fire source will ignite even for the lowest antifreeze concentrations tested and even when the antifreeze spray could be expected to extinguish most anticipated fire sources in a residential occupancy. The potential for this relatively small increase in heat release rate to reduce the effectiveness of residential sprinklers was investigated in Scope B. Results of the Scope B tests demonstrate no significant differences in the capability of water, 50% glycerin, and 40% propylene glycol to control the tested fire condition.

Antifreeze sprays that ignite and propagate away from the initial fire source are a significant concern. In addition to being a hazard on its own, ignition of antifreeze sprays extending away from the ignition source can significantly increase the heat release rate of the fire and, if confined, may result in an overpressurization or explosion.

Following Phase I of this project, it appeared that the ignition of antifreeze sprays was likely to be either localized close to the ignition source or the ignition would be sufficient to involve nearly the entire volume of the sprinkler spray. The results were readily characterized as a relatively minor localized ignition that could still allow for fire control or an ignition of the majority of the sprinkler spray that would significantly enhance the initial fire condition.

Tests during Scope A of Phase II showed that at certain antifreeze concentrations, intermittent ignition of the antifreeze spray could occur that extended away from the ignition source without involving the majority of the sprinkler spray. This expanded the results of the Phase I tests by showing that intermittent ignition of a portion of the sprinkler spray could occur and that ignition of the antifreeze spray could not be characterized as only localized or involving the majority of the spray.

The importance of the droplet size distributions and concentrations was apparent during the Phase II tests, particularly for tests where the sprinklers were positioned at 20 ft above the floor. In certain tests, ignition of the antifreeze solution was observed to extend away from the ignition source, but only into a discrete portion of the sprinkler spray. Thus, while the droplet size distribution in a portion of the sprinkler spray was sufficient to allow ignition, the droplet size distribution in the majority of the sprinkler spray was not sufficient to allow the fire to spread within the spray.

The droplet size distribution from sprinklers is not currently regulated and is often not characterized for commercially available sprinklers. In addition, even if a standard method were developed to characterize the droplet size distribution from each sprinkler over a range of operating pressures, the distribution could be modified by the properties of the antifreeze solution, the airflows in the enclosure, or the installed configuration of the sprinkler. Thus, for most residential sprinklers it does not currently appear to be practical to rely solely on the droplet size distribution as a means of preventing significant ignition of antifreeze sprays. Thus, the analysis below focuses on the characteristics of the antifreeze solution that may impact the potential for significant ignition of the spray.

B   Classification based on Solution Heat of Combustion

The heat of combustion of the antifreeze solution reduced by the additional mass of the water in the solution was investigated for its ability to characterize the relative contribution of various antifreeze solutions. The parameter investigated is the heat of combustion of the antifreeze in the solution normalized by the mass of the solution.

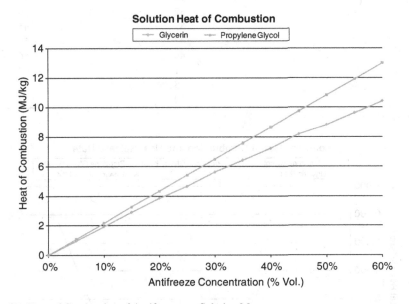

Fig. 25 Heat of Combustion of Antifreeze per Solution Mass.

Figure 25, above, shows that for a given solution concentration, a solution of propylene glycol will have a higher heat of combustion per unit of solution mass than a solution of glycerin.

The table below shows the solution concentration by volume of propylene glycol and glycerin as a function of the solution heat of combustion.

The table above indicates that a 42% solution of propylene glycol by volume has the same solution heat of combustion as a 50% solution of glycerin. Figure 26, below, compares the maximum increase in heat release rate measured in tests with k3.1, k4.9, and k7.4 pendent sprinklers as a function of the solution concentration. All data is from tests conducted with a sprinkler at 8 ft above the floor using the heptane spray burner ignition source. The k-factor of the sprinkler used in the test is indicated in the legend.

Table 6 Solution Heat of Combustion

Solution Heat of Combustion (MJ/kg)	Solution Concentration (% Vol.)	
	Propylene Glycol	Glycerin
0	0	0
1	5	6
2	9	11
3	14	17
4	18	22
5	23	28
6	28	34
7	32	39
8	37	45
9	42	50
10	46	56
11	51	62
12	55	67
13	60	73

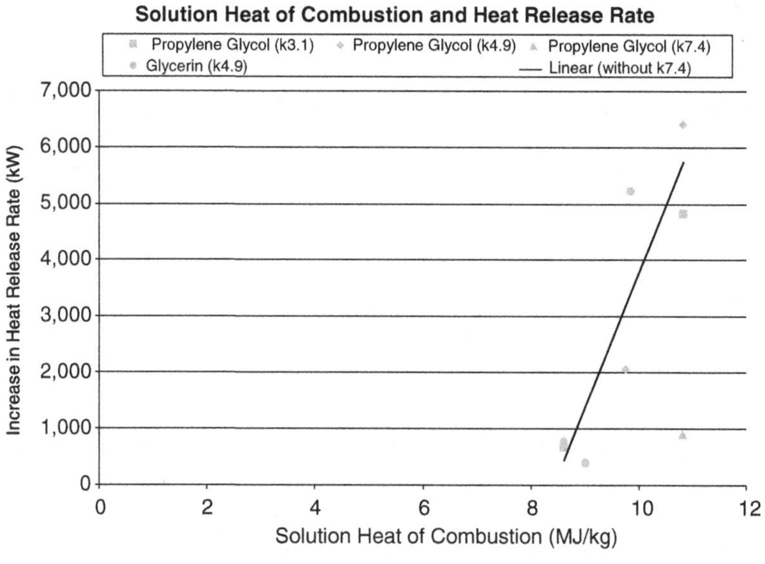

Fig. 26 Increase in Heat Release Rate as a function of Solution Heat of Combustion.

Figure 26, above, shows a strong relationship between the solution heat of combustion and the increase in heat release rate measured during tests with k3.1 and
k4.9 sprinklers. Thus, analyzing the heat of combustion of the overall solution may
be a useful method of comparing the potential contribution of different antifreeze
solutions to a fire condition. However, Figure 14 showed that the sprinkler model
also has a significant impact on the potential increase in heat release rate. Thus, the
solution heat of combustion cannot, by itself, be used to determine the potential for
ignition of an antifreeze spray.

VI
Future Research and Recommendations

The National Fire Protection Association has taken steps to implement changes based in part on information provided in prior versions of this report. Specific modifications to NFPA and other documents are not addressed in this report, but should instead be addressed through the appropriate standards revision processes. The following recommendations address research and technical areas where further investigation is warranted as a result of this work.

A Investigate the use of Antifreeze Solutions in Sprinkler Systems with Non-Residential Sprinklers

The results of this project indicate that antifreeze solutions of propylene glycol or glycerin supplied through residential sprinklers at concentrations permitted by NFPA 13 can substantially increase the heat release rate of a fire condition. The same antifreeze solutions that are used in residential sprinkler systems are also used in sprinkler systems for non-residential applications.

Residential sprinklers differ from other spray sprinklers in that they have a different spray distribution pattern. Residential sprinkler also most commonly have k-factors of 4.9, while other types of sprinklers usually have k-factors of 5.6 or greater. Thus, the droplet size distributions produced by residential sprinklers may differ from the droplet size distributions produced by other types of sprinklers in a way that impacts the potential for ignition of antifreeze sprays. Further investigation is needed to evaluate the potential for ignition of antifreeze solutions supplied by sprinkler systems not using residential sprinklers.

Based on the results of this project with residential sprinklers, research should be conducted into the use of antifreeze solutions in non-residential applications.

Code Consultants, Inc., *Antifreeze Solutions in Home Fire Sprinkler Systems,*
SpringerBriefs in Fire, DOI 10.1007/978-1-4614-3840-3_6,
© Fire Protection Research Foundation 2010

B   Characterize Droplet Size Distributions from Sprinklers

Characterization of droplet size distributions from sprinklers is an active area of research and influences the potential for ignition of antifreeze sprays. An improved characterization of droplet size distributions from sprinklers over a range of sprinkler models and operating conditions could be helpful in limiting the potential for significant ignition of antifreeze sprays. The results summarized in Figure 14 indicate that the droplet size distributions produced by certain sprinklers are less likely to create conditions where an antifreeze solution can be ignited. Further research into the droplet size distributions created by a range of sprinkler models over a range of operating conditions could be used as a basis to allow the use of antifreeze solutions under conditions when the droplet size distribution created is not anticipated to ignite.

C   Develop a Small or Medium Scale Screening Test for Antifreeze Solutions

Full scale tests were conducted as part of this project to investigate antifreeze solutions supplied by actual residential sprinklers. The tests were conducted for a wide range of operating conditions and produced a variety of droplet size distributions. Ideally, a small scale test could be used to investigate only the worst case droplet size distribution produced by residential sprinklers.

FM Global Class Number 6930, *Approval Standard for Flammability Classification of Industrial Fluids*, was identified in Phase I of this project as a test method that could be adapted to investigate the potential for ignition of antifreeze solutions. A research effort would be needed to correlate the results of any such small or medium scale test with the results of this or other full scale testing. The investigation of a small or medium scale test might also provide additional insight into the droplet size distributions that increase the potential for ignition.

D   Develop a Listing Standard for Solutions Introduced into Sprinkler Systems

NFPA 13 currently only permits glycerin or propylene glycol antifreeze solutions to be used in antifreeze sprinkler systems connected to potable water supplies. This report documents concerns with the use of certain concentrations of glycerin and propylene glycol antifreeze solution. Thus, there is a need to develop alternative solutions that could be used in instances where glycerin and propylene glycol are not suitable.

Issues of flammability, freeze protection, toxicity, and material compatibility would need to be addressed for any solutions that are introduced into sprinkler systems. A research effort would be needed to develop a series of tests that is sufficient to demonstrate the appropriateness of a new antifreeze solution and could be extended to cover other solutions that are used in sprinkler systems. The development of a listing standard that could be referenced by NFPA 13 may encourage development of alternative antifreeze solutions and help ensure that the solutions are appropriate for use in sprinkler systems.

VII
Summary

A test plan was developed for Phase II to investigate the potential for large-scale ignition of antifreeze solutions discharged from residential sprinklers. This test plan also explored the influence of antifreeze solutions on the effectiveness of residential sprinkler systems in controlling a fire condition and maintaining tenable conditions for egress.

Testing was conducted in two parts (Scope A and B). Scope A consisted of fire tests using six (6) models of sprinklers operating at pressures of 10 psi to 150 psi at elevations of eight and twenty feet. The Scope A testing was intended to investigate the potential for large-scale ignition of antifreeze sprays at pressures ranging from 10 psi to 150 psi. Scope B consisted of room fire tests, similar to UL 1626, that were designed to investigate the effectiveness of sprinklers discharging antifreeze solutions and their ability to maintain tenable conditions.

Results of the Scope A testing indicated that concentrations of propylene glycol exceeding 40% by volume and concentrations of glycerin exceeding 50% by volume have the potential to ignite when discharged through automatic sprinklers. The potential for ignition depends on several factors including the ignition source, sprinkler model, sprinkler elevation, discharge pressure, and the location of the sprinkler with respect to the ignition source. Ignition of antifreeze spray increased the measured heat release rate in certain tests with 50% propylene glycol and 55% glycerin by more than 300%. For certain test conditions, the increase in heat release rate resulting from the application of 55% glycerin solution exceeded the increase in heat release rate from the application of 50% glycerin solution by a factor of 10. A similar level of sensitivity was observed between 40% and 50% propylene glycol solutions, but not between 40% and 45% propylene glycol solutions.

The results of the Scope B testing indicated that concentrations of propylene glycol not exceeding 40% by volume and concentrations of glycerin not exceeding 50% by volume have similar performance to water as compared to the UL 1626 fire control criteria. Tests with the 40% propylene glycol and 50% glycerin solution met the UL 1626 fire control criteria and demonstrated similar performance to water throughout many of the tests.

Code Consultants, Inc., *Antifreeze Solutions in Home Fire Sprinkler Systems*,
SpringerBriefs in Fire, DOI 10.1007/978-1-4614-3840-3_7,
© Fire Protection Research Foundation 2010

The results of this research suggest that antifreeze solutions of propylene glycol exceeding 40% and glycerin exceeding 50% by volume are not appropriate for use in home fire sprinkler systems. Consideration should be given to an appropriate safety factor for concentrations of antifreeze solutions that are permitted by future editions of NFPA 13, as well as warnings and limitations outlined in antifreeze product literature. In addition, based on the flammability properties outlined in Table 4, the use of solutions of diethylene glycol and ethylene glycol in home fire sprinkler systems should also be limited.

Recommendations for further research are also provided. Further research should be conducted to investigate the use of antifreeze solutions supplied through non-residential sprinklers. The results of this study are based on tests with residential sprinklers, which are not directly applicable to other types of sprinklers due to the unique spray pattern of residential sprinklers. However, the results documented in this report are sufficient to indicate that the use of antifreeze solutions with non-residential sprinklers should also be investigated.

The droplet size distributions produced by sprinklers is an ongoing area of research that is important to understanding the potential for ignition of antifreeze sprays. Further development is needed to characterize the droplet size distributions produced by a variety of sprinklers.

The development or investigation of a small or medium scale test for ignition of antifreeze sprays may contribute to understanding the droplet size distributions of antifreeze that have the potential to ignite. Finally, the results of this research indicate that certain concentrations of glycerin and propylene glycol antifreeze solutions are not appropriate for use in residential sprinkler systems. Thus, there is a need for alternative antifreeze solutions that are not currently permitted by NFPA 13. A listing standard for antifreeze solutions or other solutions that are introduced into sprinkler systems could encourage the development of alternative antifreeze solutions and help ensure that the solutions are appropriate for use in sprinkler systems.

References

1. National Board of Fire Underwriters, *NBFU Pamphlet No. 13: Standards of the National Board of Fire Underwriters for the Installation of Sprinkler Equipments as recommended by the National Fire Protection Association.* Chicago, 1940.
2. National Fire Protection Association, *NFPA 13: Automatic Sprinkler Systems Handbook*, J. D. Lake, Ed. Quincy, MA, 2010.
3. Code Consultants, Inc., "Literature Review and Research Plan, Antifreeze Solutions in Home Fire Sprinkler Systems," 2010.
4. Underwriters Laboratories Inc., "Fire Test Data Summary for Residential Sprinklers Discharging Antifreeze Solutions," Northbrook, 2010.
5. National Fire Protection Association, *NFPA 13, Standard for the Installation of Sprinkler Systems.* Quincy, MA: National Fire Protection Association, 2010.
6. National Fire Protection Association, *Standard for the Installation of Sprinkler Systems in One- and Two-Family Dwellings and Manufactured Homes.* Quincy: National Fire Protection Association, 2010.
7. National Fire Protection Association, *NFPA 13R, Standard for the Installation of Sprinkler Systems in Residential Occupancies up to and Including Four Stories in Height.* Quincy: National Fire Protection Association, 2010.
8. National Fire Protection Association, *Standard for the Installation of Sprinkler Systems, NFPA 13.* Boston, 1953.
9. The Dow Chemical Company. (2003) A Guide to Glycol.
10. (2010, July) The Dow Chemical Company. [Online]. http://www.dow.com/glycerine/resources/physicalprop.htm
11. V. Babrauskas, *Ignition Handbook.* Issaquah, WA: Fire Science Publishers, 2003.
12. National Fire Protection Association, "Fire Hazard Properties of Flammable Liquids, Gases, and Volatile Solids," Quincy, MA, NFPA 325M, 1991.
13. S. Mannan, Ed., *Lees' Loss Prevention in the Process Industries: Hazard Identification, Assessment and Control*, 3rd ed. Burlington, MA: Elsivier, Inc., 2005.
14. M.G. Zabetakis, "Flammability Characteristics of Combustible Gases and Vapors," U.S. Dept. of the Interior, Bureau of Mines, Bulletin 627, 1965.
15. J.M. Kuchta, "Investigation of Fire and Explosion Accidents in the Chemical, Mining, and Fuel-Related Industries – A Manual," U.S. Dept. of the Interior, Bureau of Mines, Bulletin 680, 1985.
16. Noble Company. (undated) Material Safety Data Sheet: Firefighter PG Freeze Protection FluidTM Ready to Use. [Online]. http://www.noblecompany.com/Portals/0/PRODUCT%20INFO/MSDS/MSDS.FFPG.FREEZE.pdf

Code Consultants, Inc., *Antifreeze Solutions in Home Fire Sprinkler Systems,* SpringerBriefs in Fire, DOI 10.1007/978-1-4614-3840-3, © Fire Protection Research Foundation 2010

17. M. Arvidson, "An Evaluation of Anti-Freeze for Automatic Sprinkler Systems," SP Swedish National Testing and Research Institute, Boras, Sweden, Brandforsk Project 631–961, 1999.

18. J. L. de Ris, M., M. Whitbeck, and J. B. Hankins, "; K-25 Suppression Mode Sprinkler Protection for Areas Subject to Freezing," Factory Mutual Research Corporation, Technical Report J.I. 0003004619, 2000.

19. A.D. Putorti, "Simultaneous Measurements of Drop Size and Velocity in Large-Scale Sprinkler Flows Using Particle Tracking and Laser-Induced Fluorescence," National Institute of Standards and Technology, Gaithersburg, MD, 2004.

20. A.D. Putorti, T.D. Belsinger, and W.H. Twilley, "Determination of Water Spray Drop Size and Speed from a Standard Orifice, Pendent Spray Sprinkler," National Institute of Standards and Technology, Gaithersburg, MD, FR4003 Sept. 1995, rev. May 1999.

21. J.F. Widmann, "Characterization of a Residential Fire Sprinkler Using Phase Doppler Interferometry," National Institute of Standards and Technology, Gaithersburg, MD, NISTIR 6561, 2000.

22. Underwriters Laboratories, Inc., "Summary of Fire Test Data for Phase II Research of Antifreeze Solutions in Home Fire Sprinkler Systems," Northbrook, IL, September 7, 2010.

23. Vytenis Babrauskas, "Heat Release Rate," in *The SFPE Handbook of Fire Protection Engineering*, Philip J. DiNenno et al., Eds. Quincy, MA: National Fire Protection Association, 2008, ch. Section 3, pp. 1–59.

24. S. Deal, "Technical Reference Guide for FPEtool Version 3.2," National Institute of Standards and Technology, Gaithersburg, MD, NISTIR 5486-1, 1995.

25. R. L. Alpert, "Calculation of Response Time of Ceiling-Mounted Fire Detectors," *Fire Technology*, vol. 8, no. 3, pp. 181–195, 1972.

26. D. Madrzykowski and R.P. Fleming, "Residential Sprinkler Systems," in *Fire Protection Handbook 20th Edition*. Quincy, MA, United States of America: National Fire Protection Association, 2008, ch. 16, pp. 91–107.